Basic Geometry, Shapes and Patterns
2nd Grade Math
Workbook Series Vol 6

SPEEDY
PUBLISHING

Speedy Publishing LLC
40 E. Main St. #1156
Newark, DE 19711
www.speedypublishing.com

PATTERNS

What comes next?

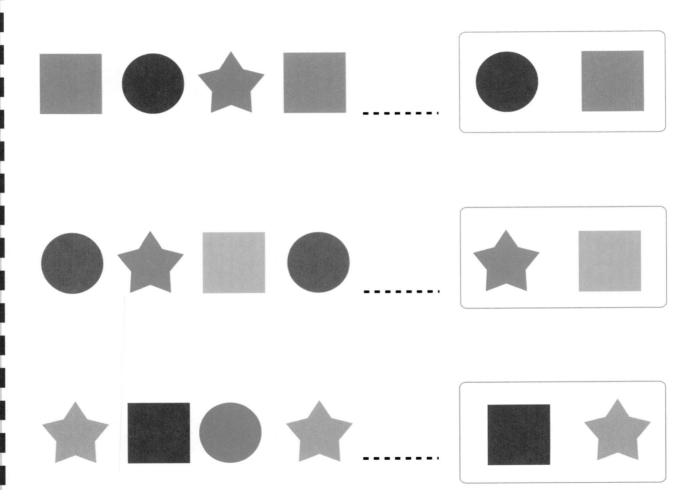

What comes next?

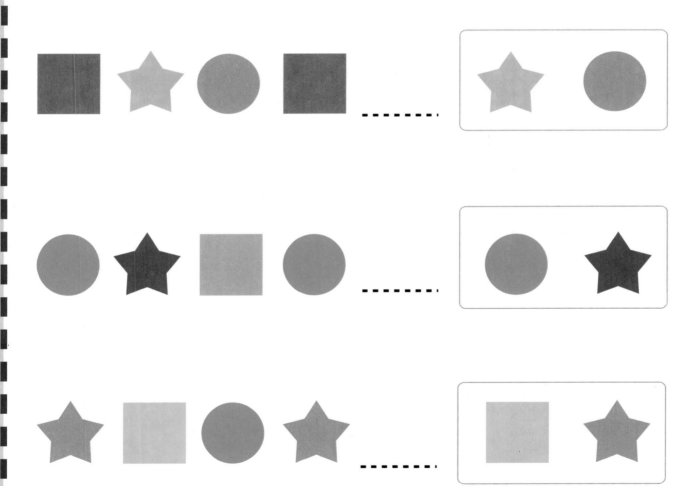

What comes next?

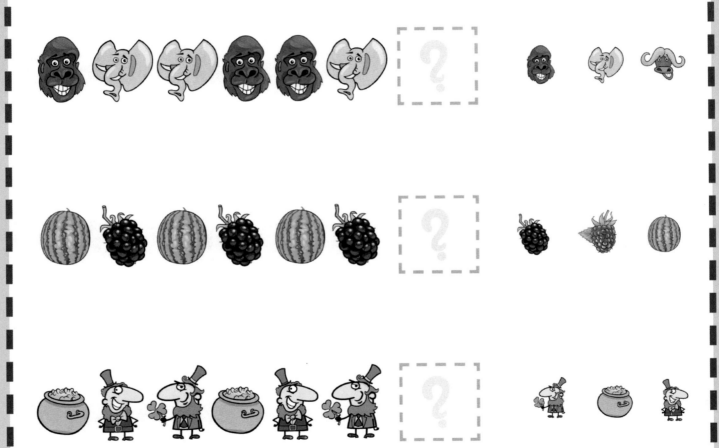

What comes next?

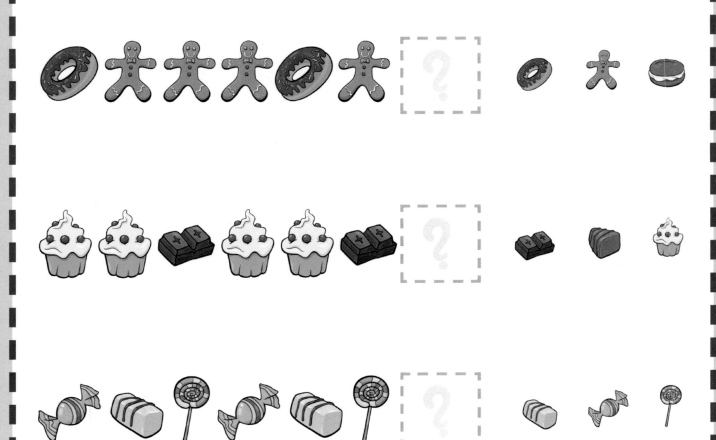

What comes next?

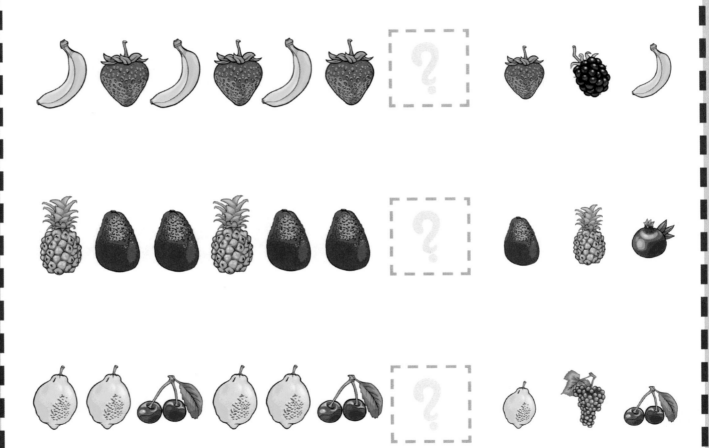

What comes next?

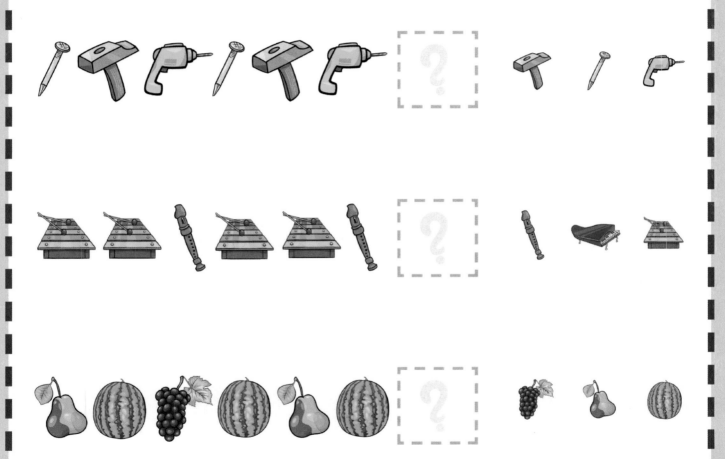

What comes next?

What comes next?

What comes next?

What comes next?

What comes next?

What comes next?

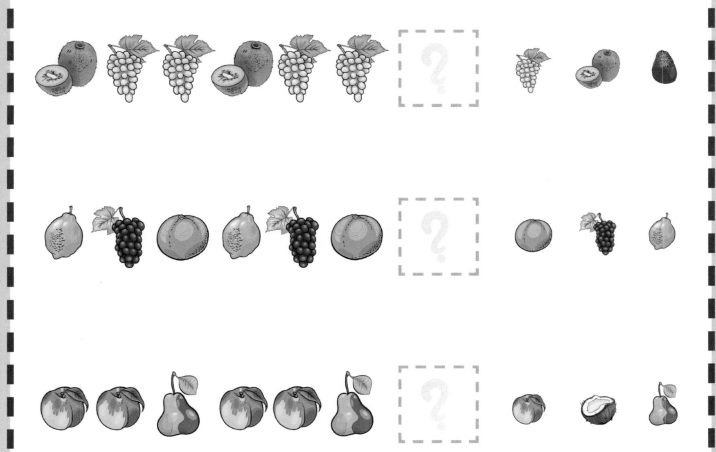

What comes next?

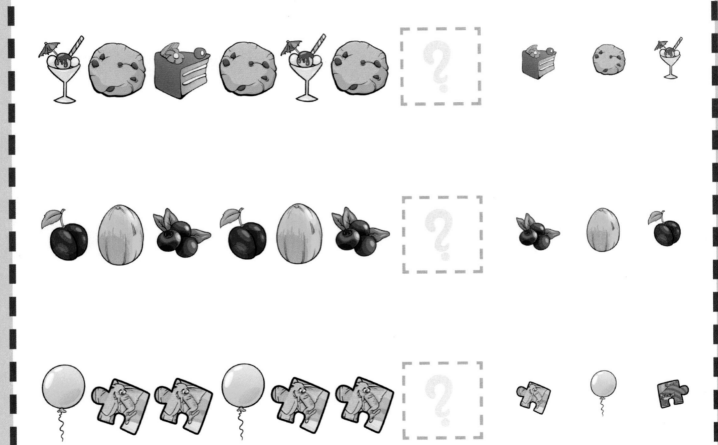

SHAPES

Many of the everyday objects with which children are familiar are SOLID SHAPES. For example, building blocks are often cubes.

Look at your surroundings, list down a cuboid you see and draw.

Look at your surroundings, list down a cone you see and draw.

Look at your surroundings, list down a cube you see and draw.

Look at your surroundings, list down a sphere you see and draw.

Look at your surroundings, list down a cylinder you see and draw.

PLANE SHAPE

is a closed, two-dimensional or flat figure.

Practice Tracing

Practice Tracing

Practice Tracing

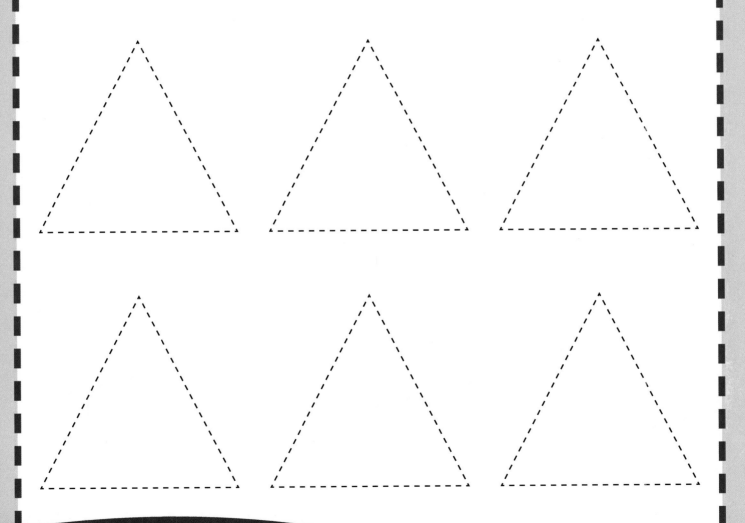

Practice Tracing

GEOMETRY

6. Draw a square with all sides having 7 cm.
   Then compute its area and perimeter.

Area = ____

Perimeter = ____

7. Draw a square with all sides having 9 cm.
   Then compute its area and perimeter.

Area = ____

Perimeter = ____

5. Draw a rectangle with 7 cm width and 5 cm length
Then compute its area and perimeter.

Area = _____

Perimeter = _____

6. Draw a rectangle with 8 cm width and 6 cm length
   Then compute its area and perimeter.

Area = ____

Perimeter = ____

2. Draw a triangle with 10 cm base, 6 cm on one side, and 8 cm on the other side.
   Then compute its area if the height is 9 cm of the triangle and perimeter.

Area = ____

Perimeter = ____

3. Draw a triangle with 3 cm base, 4 cm on one side, and 5 cm on the other side.
Then compute its area if the height is 4 cm of the triangle and perimeter.

Area = ____

Perimeter = ____

3. Draw a circle with 8 cm radius.
   Then compute its circumference and diameter.

Circumference = _____

Diameter = _____

4. Draw a circle with 5 cm radius.
   Then compute its circumference and diameter.

Circumference = ____

Diameter = ____

ANSWERS

1. A = 49 cm
   P = 28 cm

2. A = 81 cm
   P = 36 cm

3. A = 35 cm
   P = 24 cm

4. A = 48 cm
   P = 28 cm

5. A = 45 cm
   P = 24 cm

6. A = 6 cm
   P = 12 cm

7. C = 50.27 cm
   D = 16 cm

8. C = 31.42 cm
   D = 10 cm

Made in United States
Orlando, FL
19 April 2022